A SMARTER FARM:

How AI is Revolutionizing the Future of Agriculture

JACK ULDRICH

A Smarter Farm:

How AI is Revolutionizing the Future of Agriculture

Jack Uldrich

ISBN: 979-8-35093-173-0

INTRODUCTION

Artificial intelligence is, arguably, the most important technological development to happen to agriculture since the invention of electricity. And just as electricity changed the nature of work and labor by freeing up people from time-consuming and tedious jobs, AI holds the same potential to transform farming. Furthermore, just as electricity was instrumental in the creation of everything from refrigeration and power tools to the computer and the Internet, AI will similarly facilitate the creation of entirely new products, services, and industries.

The AI "revolution" will not be completed anytime soon, however. In fact, it has barely even begun. Nevertheless, it is important for farmers and agribusinesses alike to understand the basics of the technology and its trajectory for the simple reason that AI will touch upon virtually every aspect of farming.

The truth is this: The time to begin getting up to speed on artificial intelligence is *now*. Those who risk ignoring it face the prospect of being placed at a serious disadvantage to those who familiarize themselves and, ultimately, embrace the technology.

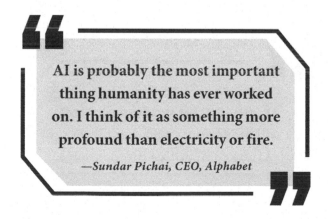

> **AI is probably the most important thing humanity has ever worked on. I think of it as something more profound than electricity or fire.**
>
> *—Sundar Pichai, CEO, Alphabet*

OBJECTIVES

The goal of this book is to serve as a primer on the topic of artificial intelligence and not only help farmers understand how AI is going to transform farming but also offer practical advice on how they can begin thinking about and leveraging the technology to become better, smarter, faster, and more efficient in virtually every aspect of their business.

Whether one thinks that the future of farming lies in increasing yields and being better stewards of the environment or that the future lies in other technologies such robotics, drones, sensors, genomics, gene editing, satellite imagery, or data analytics, AI will play an integral role in facilitating advances in all of the aforementioned areas.

By the end of this short book, readers will have an understanding of how farmers and agribusinesses are already harnessing the power of artificial intelligence and how they might do the same.

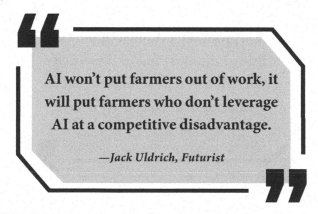

AI won't put farmers out of work, it will put farmers who don't leverage AI at a competitive disadvantage.

—*Jack Uldrich, Futurist*

WHAT THIS BOOK IS NOT

This book is neither a comprehensive overview of every company applying artificial intelligence to farming, nor will it offer an exploration of the profound moral, social, political, and ethical implications of artificial intelligence. A focus on the former would get unwieldy as there are currently an estimated 500-plus companies either working on or applying AI to farming, and the latter is beyond the scope of the author's expertise. (Note: The societal implications of AI are extremely important, but a focus on these complex issues would take away from the primary purpose of the book, which is to help farmers understand how AI will transform farming in the near future and what they should be doing about it today.)

AI:
THE FOURTH
AGRICULTURAL
REVOLUTION

The first agricultural revolution (about 12,000 years ago) marked the transition from nomadic hunting and gathering to settled agriculture. People began to domesticate plants and animals, which allowed them to settle down and form communities. This was one of the most significant transitions in human history as it led to the development of permanent settlements and, eventually, civilizations.

During the 18th and 19th centuries, the second agricultural revolution coincided with the Industrial Revolution in Europe and America, and it involved significant improvements in farming techniques and equipment. Key developments included crop rotation, better horse-drawn plows, the enhancement of seed varieties, and, eventually, the invention of mechanized equipment such as the tractor.

During this period, agricultural practices became more scientific and efficient, and the advancements allowed farmers to greatly increase their productivity, which, in turn, supported rapid population growth and urbanization. For the first time, it became possible for a small number of farmers to produce enough food to feed a large non-farming population.

The third agricultural revolution—sometimes also referred to as the green revolution—refers to a series of research, development, and technology transfer initiatives that occurred between the 1940s and the 1970s. It involved the development of high-yielding varieties of cereal grains (especially wheat and rice), the expansion of irrigation infrastructure, the modernization of farm management techniques, the distribution of hybridized seeds, and the widespread use of synthetic fertilizers and pesticides. This "revolution" dramatically increased agricultural production around the world, particularly in developing countries, and it is credited with saving hundreds of millions of people from starvation.

The integration of digital technologies into agriculture—including AI, machine learning, big data analytics, drones, sensors, precision farming, and automation—is now dramatically reshaping farming. Many experts are calling this "The Fourth Agricultural Revolution" or "Agriculture 4.0" because these technologies now promise a future where every aspect of farming is optimized—from the amount of water and fertilizer used on each plant to the timing of planting and harvesting.

The fourth agricultural revolution, as conceptualized, holds the promise of making agriculture more sustainable and efficient, and AI can be thought of as the catalyst driving this revolutionary transformation.

WHAT IS ARTIFICIAL INTELLIGENCE?

In its simplest form, AI (Artificial Intelligence) can be described as the capability of a machine to imitate intelligent human behavior. More specifically, it refers to the development of computer systems that can perform tasks that typically require human intelligence. It involves creating algorithms and models that enable machines to learn from experience, reason, and make decisions, simulating humanlike intelligence.

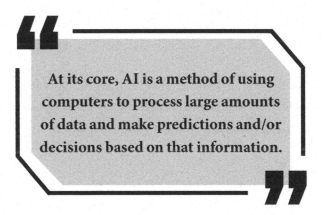

> At its core, AI is a method of using computers to process large amounts of data and make predictions and/or decisions based on that information.

Farmers, of course, have been studying and learning about the land, animals, and nutrients for thousands of years. All this knowledge was sought for the purpose of feeding people, increasing yield, and becoming more productive and profitable. What is unique about AI is the speed with which it can make sense of massive amounts of data. In an era where everything from satellites in the sky and sensors in the field to facial recognition technology in pens and pastures is growing exponentially, AI can be thought of as a way of augmenting human intelligence.

In this sense, what is unique about AI is that it will tell farmers things about their land, crops, animals, equipment, customers, and supply chains that they don't already know.

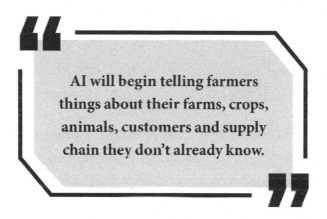

> AI will begin telling farmers things about their farms, crops, animals, customers and supply chain they don't already know.

The field of artificial intelligence is broad. Other common terms include artificial general intelligence, machine learning, deep learning, and reinforcement learning.

Artificial general intelligence or AGI, also known as "strong AI," is a higher level of AI. It refers to machines or systems that possess the ability to understand, learn, and apply knowledge across a wide range of tasks—similar to how a human thinks and acts, only exponentially quicker.

AGI is not limited to specific tasks and can perform various intellectual activities. Unlike narrow or specialized artificial intelligence, which might automate a specific task, AGI is essentially a machine with the ability to apply intelligence broadly across diverse tasks rather than being limited to a specific function.

For the farmer, AGI is what can be expected to take automation to the next level and finally deliver on the promise of precision agriculture by helping manage multiple tasks—such as planting, pest control, harvesting, and logistics—across a farm.

A third type of AI is known as machine learning. Machine learning is a subset of AI that focuses on enabling machines to learn from data without being explicitly programmed. Instead of following predefined rules, machine learning algorithms use patterns in the data to improve performance over time. It is like teaching a computer to recognize patterns and make predictions based on examples.

Farmers often face the challenge of identifying and managing crop diseases before they cause significant damage. Machine learning can be employed to create a system that detects and diagnoses diseases in crops, allowing for timely intervention.

Another area where machine learning is proving helpful is in detecting leaks in irrigation systems. Machine learning models can be trained to recognize specific and unique signatures of leaks—such as changes in water pressure or water flow—in real time.

A fourth type of AI is known as deep learning. This is a specific type of machine learning that uses artificial neural networks, inspired by the human brain's structure, to perform tasks. These networks are composed of layers of interconnected nodes (neurons) that process and transform data. Deep learning has shown exceptional capabilities in tasks like image and speech recognition and natural language processing.

Weeds, for example, are a common problem in agriculture, and their manual detection and removal can be time-consuming and labor-intensive. Deep learning networks can be utilized to automate the process of weed detection and removal by collecting data from the field (from tractors or drones). A model can then be created to recognize and detect certain weeds and this model can then be uploaded into a software system deployed on a smart tractor or a robot to identify and target the troublesome weeds.

Tip #1: One simple way to stay on top of advances in AI is to begin using it. Signing up for ChatGPT is quick and simple. Visit chat.openai.com for more information.

Tip #2: If you are wondering how AI can help farmers, you can start by asking ChatGPT what government programs, loans, or grants they may be eligible for.

Tip #3: ChatGPT is not the only AI agent available for public use. PI, Bard, and Anthropic are additional AI platforms worth investigating.

The final type of AI is reinforcement learning. This is a type of machine learning where an agent learns by interacting with an environment. The agent receives feedback in the form of rewards or penalties based on its actions. The goal is for the agent to maximize the cumulative reward over time, leading to learning and improvement in decision-making.

Farmers and agribusinesses can leverage reinforcement learning techniques to optimize their operations and improve yields, efficiency, and profitability. For example, in the area of crop management, farmers can use reinforcement learning to determine the optimal planting schedule, irrigation strategy, and fertilizer application for different crops. The AI agent does this by learning from historical data and real-time observations and then adapting its decisions based on weather patterns, soil conditions, and crop health to maximize yields while also minimizing resource usage.

When combined with advances in robotic technology, reinforcement learning might prove to be particularly useful in enhancing the performance of agricultural robots used for harvesting. In such a case, the robot could learn to identify ripe fruits and optimize harvesting techniques to both reduce labor costs and increase harvest efficiency.

Tip #4: Think of AI as a new farm-hand if you will. The only difference is that this worker will grow exponentially smarter over time and can work around the clock.

Tip #5: If you are from a large farm or an agribusiness, it is not too soon to begin shopping around for a "chief AI officer." This position will require a deep understanding of AI and how it might be applied to all aspects of your business.

HOW AI ALGORITHMS PROCESS AND LEARN FROM DATA

A I algorithms process and learn from examples through a method called supervised learning. In this approach, the algorithm is presented with a dataset containing input-output pairs, where the inputs are examples and the outputs are corresponding labels or targets. The algorithm's goal is to learn a mapping between the inputs and the outputs so that it can make accurate predictions on new, unseen data.

Using farming and agriculture as an example, the first thing that needs to be done is to collect a dataset containing examples of various farming practices and their corresponding outcomes. Such a dataset may include information about soil types, weather conditions, irrigation methods, crop types, and the yield obtained for each combination of factors.

Next, the collected data needs to be preprocessed to ensure that it is in a format suitable for the AI algorithm. This may involve normalizing numerical values, encoding categorical variables, and handling missing data. (This is why cleaning up data is often the first thing that needs to be done before any farmer or agribusiness can begin to harness the benefits of AI.)

Following this, the third step is to choose an appropriate AI model for the task at hand. For agricultural applications, popular choices include decision trees, support vector machines, or more advanced models like neural networks.

The AI model is then trained on the preprocessed dataset. During training, the model uses the input-output pairs to adjust its internal parameters and learns to make accurate predictions. In the context of agriculture, the model aims to understand the relationships between different farming practices and their impact on crop yields.

Once the model is trained, it is evaluated on a separate dataset. This dataset is called the validation or test set and it is used to assess the performance of the model. The performance metrics, such as accuracy, indicate how well the model generalizes to new, unseen examples.

After the model is trained and evaluated, it can then be used for making predictions on new data. For example, if a farmer wants to optimize their crop yield, they can input information about their current farming practices (e.g., soil types, weather conditions, irrigation methods, etc.), and the AI model can predict the expected yield based on the patterns it learned during training.

As the AI model receives more data and examples from different farms, it can continuously improve its predictions and become more accurate over time. The more diverse and relevant data it has, the better it can generalize to different farming scenarios and provide valuable insights for farmers to make informed decisions and optimize their agricultural practices.

Using a hypothetical example, let's consider the case of Steve, who grows potatoes on his farm in eastern Washington and wants to optimize his yield.

Steve will begin by deciding what data to collect about his farming practices, environmental conditions, and the corresponding potato yields over several seasons.

Next, he will set up a data collection system on his farm to gather relevant information during each growing season, including the following:

- Soil characteristics: pH level, nutrient levels (like nitrogen, phosphorus, and potassium), moisture content, and soil texture (sand, silt, and clay)
- Environmental conditions: Temperature (at various times of the day), humidity levels in the air and the soil, wind speeds, solar radiation levels, and sunlight
- Farming practices: Irrigation methods (type and frequency), fertilizer application (types and quantity), and pesticide usage
- Plant-related data: Planting dates, plant growth stage, plant height, leaf area, disease incidence, yield, quality, and harvest dates

After collecting the data, Steve might realize that some of the data points have missing values due to sensor failures or human errors, and he will strive to clean up his data. This will include finding outliers in the data, which may have been caused by anomalies or measurement errors.

Once this task is completed, Steve will choose an appropriate machine learning algorithm suitable for his situation. In his case, a model that can handle both numerical and categorical functions capable of capturing complex relationships between multiple variables may be most appropriate.

With his cleaned-up dataset, Steve will then split the data into a training set and a validation set. He will use the training set to train the model on historical data, where the input features are the various factors related to his farming practices, and the output is the yield obtained during each season.

After training the model, Steve will evaluate its performance on the validation set. Once satisfied with the model's performance, Steve will deploy it on his farm and then, whenever he wishes to estimate the yield for a new growing season, he will simply input the relevant data about the soil, environment, and farming practices into the AI model, and the model will predict the expected yield based on the patterns it learned during training.

Over time, as Steve collects more data from each season and feeds it into his AI model, the model can continuously learn and improve its predictions. This allows Steve to make data-driven decisions about his farming practices while also striving to optimize resource allocation and maximize his potato yield.

WHY IS ARTIFICIAL INTELLIGENCE ACCELERATING THE FUTURE?

In 2014, a group of scientists from Los Alamos National Laboratory founded Descartes Labs for the purpose of processing satellite imagery to help farmers, commodity traders, and governments make more accurate predictions about global food supplies as well as detect early warning signs of famine. Since that time, there have been four developments which have contributed significantly to processing satellite imagery.

First, the resolution on satellite cameras continues to improve. Today, the resolution on some satellites is so good that they can detect the slightest coloration in a plant's leaves. This improved resolution is helping farmers identify diseases and pests much sooner.

Second, the cost of putting satellites in space has plummeted. In 2012, a "blue ribbon" committee commissioned by the president of the United States estimated that a high-end rocket capable of launching satellites into space would take twelve years to manufacture and cost $36 billion. Elon Musk and his company, SpaceX, accomplished the task in 6 years—50 percent faster than the experts had predicted. Moreover, he did it for less than $1 billion—a mere fraction of what the committee had expected.

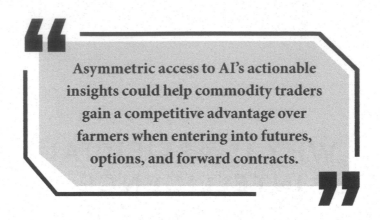

> Asymmetric access to AI's actionable insights could help commodity traders gain a competitive advantage over farmers when entering into futures, options, and forward contracts.

Over the past few years, SpaceX has launched hundreds of satellites into space, and while not all of the satellites collect data on land, crops, and water, a number of them do. This, in turn, has led to an explosion in the amount of data available to farmers. Today, petabytes of data are collected on everything from the growth of specific crops to the growth of cities. (The amount of satellite data available will continue to grow at a near exponential rate as the number of satellite launches grows into thousands and then tens of thousands.)

Third, the cost of storing, processing, and accessing all of this data has also plummeted. Today, Google Cloud, Amazon Web Services, and Microsoft Azure are among the world's leading cloud providers. In the case of Descartes Labs, the company has migrated its core information technology infrastructure to Amazon's platform, allowing it to affordably store vast amounts of geospatial data.

Artificial intelligence is now the fourth piece of the puzzle, and Descartes's algorithms and machine learning models have continued to improve to the point where the company's goal of providing the world's 600 million farmers with helpful information—something that can boost their yields and lower their input costs by identifying which crops are unhealthy and/or under stress—is within the realm of possibility.

One other aspect farmers would want to consider is how commodity trad-ers are already accessing Descartes Labs' information to make better and more informed buying and selling decisions. This asymmetric access to AI's actionable insights could lead to a competitive disadvantage for farmers when entering into futures, options, and forward contracts.

Case Study:
Carbon Robotics

Carbon Robotics is a Seattle-based start-up focused on developing solutions that harness the power of robotics to automate tasks for farms of all sizes. One of the most tedious, time-consuming, and challenging tasks in farming is weeding. In 2022, the company began shipping its LaserWeeder to farmers across North America. In 2023, the LaserWeeder, which can operate on 40 different crops and cover two acres per hour, was named "Best AI Solution in Agriculture" by the AI Breakthrough Award Committee.

The LaserWeeder, which can operate either autonomously or by being pulled behind a tractor, is 20 feet wide and consists of three rows of ten lasers each. After the company's proprietary AI distinguishes between a crop and a weed, it uses onboard high-resolution cameras to identify weeds and use lasers—which have millimeter accuracy—to disrupt the cellular structure of the weeds. Its technology is so effective it kills 99 percent of all weeds, including those so small that a human hand wouldn't be capable of grabbing them.

LaserWeeder can currently identify 80 different types of weeds and is capable of killing up to 200,000 weeds per hour. The machine does the equivalent work of 70 people, can work day and night, and can reduce the cost of weed control by 80 percent. (This combination of cost savings allowed one farmer to pay for the technology within a single year.)

> **The AI-enhanced LaserWeeder is capable of killing 200,000 weeds per hour.**

As an added benefit, Carbon Robotics' LaserWeeder has been demonstrated to increase the yields of many crops because those crops are no longer being unnecessarily damaged by herbicides. Moreover, because the LaserWeeder works without harmful chemicals, not only has it received certification from leading organic organizations, but it is also proving a helpful ally in combating weeds that are becoming resistant to glyphosate.

WHY IS AI IMPORTANT
TO FARMERS?

Artificial intelligence is becoming increasingly important to farmers for a variety of other reasons. Its integration into modern agriculture—often referred to as "precision agriculture"—offers solutions to many of the challenges faced by today's farmers.

1. **Enhanced Productivity**: AI algorithms can analyze vast amounts of data quickly, helping farmers make decisions that maximize yield. This includes optimal planting times and soil management strategies.

2. **Efficient Resource Use**: AI can optimize the usage of water, fertilizers, and pesticides. For instance, by analyzing soil moisture data and weather predictions, AI can recommend precise irrigation schedules and decrease water use.

3. **Labor Reduction**: With a declining agricultural workforce in many areas, AI-driven machines can fill the gap, automating tasks such as planting, harvesting, and sorting.

4. **Financial Decision Support**: AI can help farmers make more informed financial decisions by predicting market demand and price fluctuations and recommending optimal times to sell crops.

5. **Data-Driven Insights**: With the help of sensors and IoT (Internet of Things) devices, AI can process enormous amounts

of real-time data, providing insights that were previously impossible or highly time-consuming to obtain.

6. **Customized Solutions**: Every farm and field is unique. AI can help tailor solutions specific to the individual needs of a farm or even a particular section of a field instead of adopting a one-size-fits-all approach.

At a more specific level, AI will also assist in and improve the following:

1. **Precision Agriculture**: AI-powered systems will analyze data from sensors, satellites, and other sources to optimize crop management. Farmers can then use these insights to make better informed decisions on what crops to plant and when to plant them. AI will also assist farmers in making more informed decisions regarding irrigation, fertilization, and pest control.

2. **Crop Monitoring**: Drones, cameras, and other AI-enabled imaging technologies will be used to monitor crops' health and detect early signs of diseases or nutrient deficiencies. AI can analyze soil samples to determine its health, composition, and needs, thereby helping farmers tailor their practices to maintain or enhance soil health. This will allow farmers to take timely actions and prevent potential loss of crops.

3. **Yield Prediction**: Machine learning algorithms will continue to get better over time at predicting crop yields based on historical and real-time data. This will help farmers make more informed planting and harvesting decisions.

4. **Pest and Disease Detection**: AI is already assisting farmers in identifying a growing number of pests and diseases affecting crops more accurately. A partial list of plant diseases includes late blight, powdery mildew, citrus canker, and fire blight. Some of the pests AI has proven effective at identifying early include

fall armyworm, aphids, whiteflies, and fruit flies. AI will enable farmers to use targeted interventions and reduce the need for broad-spectrum pesticides. The latter will help promote more sustainable practices.

5. **Supply Chain Management**: AI can optimize logistics and supply chain management by creating better demand prediction models, optimizing the routes of crop shipments, monitoring shipments in real time, assisting with predictive maintenance, improving quality control, providing smarter consumer insights, and improving traceability.

6. **Climate Adaptation**: With AI's help, farmers can analyze climate patterns and trends and make more informed decisions on which crops to plant and when to plant them. One area where AI is likely to play an increasingly important role is the area of risk management. As AI-driven weather forecasting models improve, farmers will receive more accurate and localized predictions of weather patterns, extreme events, and climate-related risks. They can then use this information to plan their planting, irrigation, and harvesting schedules, as well as take precautionary measures to protect their crops and livestock from weather-related threats.

THE LOWEST HANGING FRUIT: USING AI TO GAIN ACCESS TO AGRICULTURAL KNOWLEDGE AND EXPERTISE

Artificial intelligence is a complex technology and it can be intimidating. In addition to experimenting and playing around with AI by using such platforms as ChatGPT or Pi, the next step farmers can take is to investigate other AI-powered platforms and mobile applications that can provide them with easy-to-access agricultural knowledge and expertise. A number of AI platforms offer personalized recommendations, and expert advice on crop management, pest control, and soil health. Small farmers can leverage these resources to enhance their farming skills and implement effective strategies for increased yields.

Plantix, for example, is a mobile app that utilizes AI and machine learning algorithms for crop disease and pest identification. It was created by a husband–wife duo who had a mission to make agriculture more sustainable while increasing yields and income for the tens of millions of small farmers around the world who cannot afford to hire a professional agronomist.

Using the app, farmers can take photos of affected plants, and its AI algorithms can analyze the images in real time to identify specific diseases and pests. Plantix then provides recommendations for treatment and management practices based on the identification.

At the time of this publication, Plantix had over 5 million photos and could accurately identify over 350 different diseases. The company recently entered into a partnership with BASF and is using an additional $20 million investment to improve its rapidly growing database.

Case Study:
Plantix and Green Acres Farm

Green Acres Farm is a small-scale, family-owned farm located in a rural area of India. The farm cultivates rice, wheat, and vegetables to sustain the livelihood of the farmer's family as well as generate some surplus for the local market.

Its challenge lay in identifying and managing crop diseases effectively. Traditional methods of disease diagnosis were time-consuming and often required expert assistance, which was not readily available in the remote farming community.

Green Acres Farm learned about the Plantix mobile app through an agriculture extension officer's recommendation. Intrigued by the idea of using AI and image recognition technology to diagnose crop diseases, the farmer decided to give it a try.

After downloading the Plantix app from his smartphone's app store and installing it for free, the farmer used the app to capture images of plants that were displaying unusual symptoms. After analyzing the images, the app provided a list of potential diseases that might be affecting the plants, along with detailed descriptions and reference images for comparison.

Based on the diagnosed diseases, Plantix offered personalized treatment recommendations and management strategies. This included suggestions on appropriate pesticides, fungicides, or nutrient applications.

Green Acres Farms also became part of the Plantix community and connected with other farmers, shared its experiences, and learned from others facing similar issues.

With the help of the Plantix app, Green Acres Farm was able to do the following:

- Detect plant diseases at an early stage, enabling quick action and preventing the spread of diseases to other parts of the farm
- Save money by using targeted treatments recommended by the app and eliminating unnecessary pesticide and fertilizer usage
- Improve the overall health and productivity of the crops with timely and accurate disease management, resulting in better yields

Plantix can be downloaded for free on either Google or Apple App stores and it is easy to use. Simply click a picture to find out what's wrong with the plant and how to treat it. The app will even provide both conventional and organic treatment options. Better yet, as more farmers use the app and add to Plantix's growing library of images, the system will get more intelligent over time.

By allowing farmers to geotag diseases, the app is also able to track the movement of pests and diseases in real time. An additional benefit is that the app is helping combat the growing problem of insecticide resistance by helping farmers apply treatments only where and when necessary.

Other chatbots worth reviewing and exploring include WeFarm, AgroPedia, AgriBot, Farm.Ink, Ava, and, most recently, Ask Norm.

In April 2023, the Farm Business Network (FBN) announced the release of the agricultural industry's first AI-powered agronomic advisor dubbed "Norm"—in honor of Dr. Norman Borlaug, the "father of the green revolution" and Nobel-prize-winning agronomist, who is credited with feeding billions of people around the globe. FBN is inviting its members to experiment with the platform, which looks and operates in a manner similar to ChatGPT.

> **With Farmers having to make hundreds of decisions each season on topics as diverse as soil chemistry, plant genetics, chemical applications, equipment repair, and commodity price hedging . . . Farm Business Network envisions a future where 'Norm' can serve as a frontline advisor.**

At the present time, Norm is only available to FBN members. While it is not a substitute for a trained, professional agronomist, the platform is getting smarter over time. In addition to learning from the queries the bot cannot currently answer, FBN's data science team is working to make it smarter by supplementing answers and feeding it the latest knowledge from the USDA, various agronomy publications, Agricultural extension services, and other leading industry sources.

Some questions a farmer can already ask Norm are as follows:

- Based on the soil samples I've provided from Field X, what would be the optimum fertilizer blend to achieve the highest yields for my corn crop?

- Given the forecasted weather patterns for the upcoming month in southwestern Minnesota, what's the best window for planting soybeans to ensure germination success?

- Looking at my farm's historical yield data and the latest disease alerts, should I be concerned about any specific crop diseases this season? If so, which fungicides have been proven most effective for treating those diseases?

- Given the current market prices and my farm's production costs, when would be the most economically advantageous time to sell my wheat crop over the next three months?

- How do the growth patterns of the hybrid canola seeds I've recently purchased compare to the previous ones I've used and based on this knowledge, should I make any adjustments in my sowing density or irrigation schedule?

- At what temperatures do corn rootworms begin to lay eggs?

At a more generic level, here are a few additional areas where Norm may prove useful:

- **Chemical intelligence**: Norm can help identify generic alternatives, application rates, pest and disease targets, tank mix, and nozzle suggestions.

- **Input Guidance**: Norm can provide guidance on the best seed varieties based on geographic region, soil type, and climate.

- **Pest and disease strategies**: Norm can help farmers develop strategies for dealing with common pest and disease issues and incorporating practice management with specific products. Specifically, it can recommend which herbicides, pesticides, and fungicides will be most effective on a given field.

- **General agronomic advice**: Norm can provide insights on everything from irrigation to fertilization to crop rotation, providing tailored answers to farmers based on their specific circumstances.

- **Livestock and animal health**: Norm can help livestock producers better understand the diseases affecting their animals and the pharmaceuticals that may prove most effective in treating those ailments.

WHY EVEN SKEPTICAL FARMERS SHOULD CARE ABOUT AI

Even if you are not inclined to use artificial intelligence on your farm or in your operations, there are still a number of reasons why you would want to know about AI and follow advances in the field.

First, you may already be using AI without even knowing it. The easiest application to understand is weather prediction. Over the past years, weather forecasts have gotten significantly better and more localized. In addition to better satellite images and data, AI algorithms have helped create better and more accurate forecasts. This has helped minimize the risk of crop failure.

Second, many AI applications are either free or low-cost. The case study of Plantix and Norm offered earlier are but two examples. Download the apps and begin playing around with them.

Third, over time, AI can be expected to have a significant impact on export markets and the price of certain crops. It is currently estimated that there are approximately 600 million farmers worldwide. Most of these farmers own or work on small plots. Most countries around the world have some sort of AI initiative aimed at helping farmers become more efficient, productive, and profitable.

> AI is changing business models. Soon 'drones-as-a-service' utilizing AI could replace expensive chemical applicators.

For instance, China has created an "AI Smart Farm" initiative as part of its latest 5-year plan and it's working to integrate the sensors, drones, data analytics, blockchain, and robotics with AI to help its farmers increase yield, decrease input costs, and detect early signs of disease and pests.

In India, CropIn is helping small farmers use satellite data to monitor crops in real time, and in Brazil, AI-driven robots are helping address labor shortage issues. In Africa, farmers are using AI-driven applications to process data about market pricing so they can make smarter decisions about when and where to bring certain crops to market.

Fourth, AI will continue to accelerate the transformation of the agriculture industry. By virtue of its ability to design new proteins and enzymes, AI may facilitate advances across a variety of farming fields. AI, in combination with continued advances in genomics and gene-editing technology, may further speed new seeds into market.

Fifth, AI may play an indirect role in creating new business models. One example is Rantizo, which is using swarms of drones to apply various chemical inputs to a farmer's field. AI is the enabling technology that allows these drones to operate in partnership with one another. In the near future, farmers may not need to make sizable investments in chemical applicators and may instead utilize "drones-as-a-service."

Sixth, while not all farmers today not care about their carbon footprint, this could change as consumers become better able to track how and where their food was grown. Those farmers who are not using AI to document how they are working to lower their carbon footprint may find themselves on the wrong side of consumer opinion and could receive a lower price for their crop as a result.

Lastly, if a farmer doesn't utilize AI, he may find himself at a competitive disadvantage to those farmers and commodity traders who are using the technology to make better, smarter, faster, and more profitable decisions.

KEY INDUSTRY CHALLENGES IN AGRICULTURE

Today's farmers are tasked with feeding the planet by growing healthier and more productive food and feed, while at the same time serving as responsible stewards of the land and its resources. To better understand the magnitude of the challenge facing the industry, here are the five major challenges confronting the world's farmers:

1. A growing worldwide population: According to the United Nations, the number of people living on the planet is expected to reach 8.6 billion by 2030 and almost 10 billion by the year 2050. Put another way, the world will be adding about 83 million new people every year. This is the equivalent of adding a country the size of Germany to the world's population every year for the next three decades. In aggregate, this growth will represent an increase of 70 percent in the amount of food farmers need to produce. This percentage could grow even larger as many of the citizens of the world's underdeveloped countries—primarily in the Global South—grow more prosperous and begin demanding more calories and protein in their diets.

2. A reduction in arable land: According to the UN's Convention to Combat Desertification, the world is losing 12 million hectares of land every year. (A hectare is 10,000 square meters and

12 million hectares is roughly the size of Great Britain.) This suggests that farmable land could be halved by 2050 due to soil degradation, erosion, water scarcity, rising sea levels, and the growing desire to return some land to its natural state as a forest or prairie.

> **The world is adding the population equivalent of a new country the size of Germany every year while, at the same time, losing an amount of farmable land equal in size to Great Britain every year.**

3. <u>A shrinking workforce</u>: The number of people working in agriculture has fallen rapidly—from almost 40 percent of the global workforce in 2000 to 27 percent in 2020. By 2050, the percentage is likely to drop even further, meaning that fewer farmers will be responsible for growing almost twice the amount of food. The primary causes of this shrinking workforce are land consolidation, urbanization, and an aging farm population.

4. <u>An increase in climate-related disruptions</u>: From extreme droughts and hurricanes to wildfires, windstorms, and floods, the number and severity of "extreme weather events" is increasing, and these events are expected to lead to decreased crop productivity. It has been estimated that an increase of even 1°C in the global average temperature could cut the yields of the world's leading caloric crops, including wheat, maize/corn, and rice, by 3 to 6 percent.

5. Pests: According to the Food and Agricultural Organization (FAO), somewhere between 20 to 40 percent of all crops worldwide is lost to pests, plant pathogens, and weeds. Damages from plant diseases alone are estimated to total $220 billion per year. They mostly affect wheat, maize/corn, rice, potatoes, soybeans, and cotton.

In simple terms, the confluence of these challenges suggests that the agriculture industry will need to feed far more people with fewer farmers while also using less land and fewer resources in a rapidly changing and increasingly complex and chaotic environment. It is difficult to see how farmers will meet these grand challenges without harnessing and leveraging the power of artificial intelligence.

Case Study:
How AI is Helping Reduce
Food Wastage Afresh

Approximately 40 percent of all food grown is wasted. In the U.S., it is further estimated that supermarkets alone are responsible for almost 10 percent of all food waste. The annual cost of growing, transporting, storing, and dispensing food that is never eaten is $218 billion in the U.S. alone.

Afresh, a San Francisco-based start-up, is using AI to assist stores in forecasting sales and optimizing orders to minimize food waste. According to the company, in the 300 stores where its AI platform is being utilized, it is reducing industry "shrink" (the industry term for food waste) by 10 percent. Other benefits of applying AI include adding an average of 2 days to the post-sale shelf life of many products, increasing labor efficiency by 20 percent, reducing stockout reduction by 80 percent, and increasing sales by 3 percent. As of late 2023, Afresh is responsible for eliminating 7 million pounds of food waste, avoiding 140 million gallons of water usage, and removing 62,000 tons of carbon dioxide (CO_2) out of the atmosphere.

In the grand scheme of things, these figures are a proverbial drop in the bucket. In the very near future, however, every farmer should be able to leverage AI to improve predictive analytics to tell them the optimal time for planting and harvesting. And at the consumer level, AI—in the form of smart refrigerators—might soon suggest recipes that will use items in hand or those which are about to reach their expiration date. (In fact, in 2024, Samsung hopes to introduce "Vision AI" to allow consumers to do exactly this.)

AI AND PLANT/
CROP BREEDING

For generations, humans have selected the best or most promising plants to breed in order to produce crops with favorable qualities. These traditional breeding methods, although effective over time, were slow and often involved a degree of unpredictability. Thanks to continued advances in computational power, cloud computing, gene sequencing, and AI (with its capacity to analyze vast amounts of genetic and environmental data), researchers can now significantly accelerate the breeding process. This precision plant breeding promises to be one of the most transformational shifts in breeding history, and it will bring new varieties of seeds to market sooner than expected.

One simple way to understand this is to consider a Rubik's Cube. For most people, the cube is a difficult, if not an almost impossible, task. The most skilled human puzzle solvers can solve a Rubik's Cube in about 4 seconds. Artificial intelligence can do it in .38 of a second—15 times faster than the best human. And AI, unlike humans, will only continue to get faster.

Now think of a corn seed—or any seed—as a vast, complicated Rubik's Cube with a multitude of attributes, each defined by genetics. AI can analyze billions of data points and can review millions of different combinations in order to rapidly identify and eliminate negative traits from the seedwhile simultaneously enhancing the most beneficial traits.

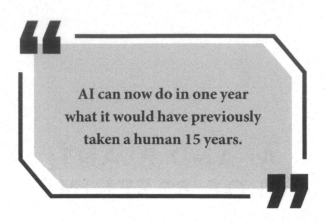

> AI can now do in one year what it would have previously taken a human 15 years.

This speed in breeding will not only translate into a quicker return on investment for agribusinesses, but it will also ensure farmers are equipped with the best plant varieties to thrive in their specific environmental conditions. Put another way, AI is developing next-generation seeds for specific geographic locations and getting these new varieties to market sooner than was previously possible.

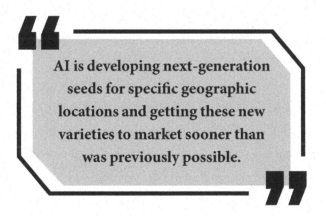

> AI is developing next-generation seeds for specific geographic locations and getting these new varieties to market sooner than was previously possible.

By pinpointing the exact genetic combinations that result in crops requiring less water or specific nutrients, AI can also help farmers adopt more sustainable practices. Additionally, optimized breeding results in healthier, more productive crops and this, in turn, allows farmers to get better market prices and increase their profit margins.

Case Study: Bayer and the Genetic Analysis of Crops

Bayer, especially after its acquisition of Monsanto, has been at the forefront of integrating AI into agriculture. The company recognizes the potential of AI in expediting and refining the plant breeding process.

Bayer and other AI leaders are no longer interested in simply selecting the most promising seed candidates during product development. Rather, they are interested in identifying the best seed to meet the unique needs of every field depending on that field's moisture level, nutrient density, and soil temperature.

This precision plant breeding is based on the reality that no two fields are the same. Climate, pests, and growing conditions vary from farm to farm, and Bayer has developed a more informed seed pipeline using advanced genomics, data science, and AI.

Bayer's extensive database of plant genomes, combined with its AI algorithms, can predict how different genetic combinations will affect a plant's yield, disease resistance, and tolerance to different environmental factors

Under Bayer's guidance and through the use of AI-driven breeding techniques, certain strains of corn have been identified and developed to show resilience to drought conditions. This has enabled farmers in areas with unpredictable rainfall or water shortage to achieve better yields.

By analyzing data from various environmental conditions and combining it with genomic data, Bayer has also developed corn varieties with increased resistance to specific pests or diseases. This means less crop loss as well as a reduced need for pesticides.

The application of AI in analyzing vast amounts of data from field trials has also enabled the quicker identification of high-yield-producing corn strains. This accelerated process ensures that farmers get access to these high-yield varieties sooner than through traditional breeding methods.

CROP MONITORING

In addition to using apps such as Norm and Plantix, farmers can also avail themselves of other AI-based systems which utilize sensors, drones, and satellite imagery to monitor crops continuously.

These systems can analyze data such as plant health, growth patterns, and nutrient levels. Moreover, by employing machine learning algorithms, they can also detect diseases, pests, and nutrient deficiencies at an early stage, allowing farmers to take timely action.

Many farmers are already using in-field sensors. In one case study, a corn farmer in Illinois used Climate Corps' Climate FieldView™ to analyze data collected from sensors installed on the farmer's equipment. The AI-driven platform helped optimize planting practices by adjusting seed population and improving spacing. As a result, the farmer achieved an yield increase of 9–12 bushels per acre, leading to significant profitability gains.

John Deere also has technologies utilizing AI and sensor-based solutions. A soybean farmer in Iowa used John Deere's Operations Center to monitor and analyze real-time data from sensors mounted on his equipment. The AI-powered platform provided insights into crop health, growth patterns, and pest infestations, enabling the farmer to make data-driven decisions. As a result, the farmer optimized inputs, reduced costs, and achieved a 5 percent increase in overall crop yield.

Drones are another increasingly popular way to collect information on crops. PrecisionHawk is an American drone technology company that offers comprehensive data solutions for agriculture, including drones equipped with multispectral and thermal sensors.

The company, which was acquired by Field in March of 2023, utilizes AI and machine learning algorithms to analyze the collected data and provide actionable insights to farmers. In one case study, PrecisionHawk partnered with Opus One Winery in Napa Valley to improve vineyard management and grape production.

By deploying drones equipped with specialized multispectral sensors to collect high-resolution aerial imagery of the vineyard, Opus One was able to capture data beyond what the human eye could see, including infrared imagery that revealed valuable information about the health and stress levels of the grapevines. Specifically, its platform analyzed the data and generated actionable insights related to vine health, disease detection, water stress, and nutrient deficiencies.

The benefits included identifying potential disease outbreaks in the vineyard, ensuring consistent grape quality for their premium wines, and maximizing productivity.

By identifying areas with water stress, the winery could also optimize their irrigation practices and ensure efficient water use. This resulted in both water savings—which is increasingly important in drought-stricken regions around the world—and improved grape quality.

Another company specializing in crop monitoring is Taranis, whose stated mission is to provide "AI-powered crop intelligence solutions" to the hundreds of farmers currently using its technology. (The company currently covers over 3 million acres.)

Like PrecisionHawk, Taranis uses drones, satellite imagery, and AI-powered analytics to monitor crops, detect early signs of pests, diseases, and nutrient deficiencies, and help farmers make data-driven decisions to optimize yields.

The company typically charges about $10–$18 per acre and uses independent gig-economy operators to fly drones over a farmer's field every 10–12 days—or about 4 to 6 passes per season. Its sub-millimeter resolution is so good it can reportedly count the dots on a ladybug's shell.

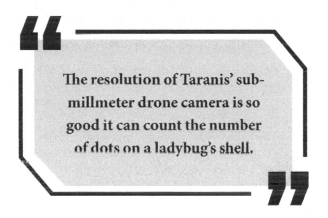

The resolution of Taranis' sub-millmeter drone camera is so good it can count the number of dots on a ladybug's shell.

In one case study involving a large-scale corn farm in the Midwest, Taranis deployed drones equipped with high-resolution cameras and advanced sensors to capture detailed imagery of the cornfields throughout the growing season.

The drones flew at low altitudes and captured data at a sub-centimeter resolution. This highly accurate data yielded valuable and actionable insights on the corn's health, stress level, pest infestations, nutrient deficiencies, and weed presence.

The most significant benefit, however, was in the area of fertilizer application. By analyzing crop health, the farmer could precisely apply fertilizers to areas that required them most. This optimized fertilizer usage minimized excessive application and reduced fertilizer costs without harming crop productivity.

Taranis' AI analysis also helped the farmer identify areas with specific pest pressures, enabling targeted pesticide application only where necessary. This reduced the farmer's overall chemical usage and improved his bottom line while also serving to better protect the environment.

As an added benefit, after the growing season, Taranis uses anonymous data to provide benchmarking comparisons with other farms in the area to help its farmers make better decisions for the upcoming planting season.

Case Study:
Trace Genomics

Trace Genomics is a company based in Ames, Iowa that provides farmers with insights into their soil's health using advanced DNA sequencing and machine learning. By analyzing soil samples, Trace Genomics helps farmers better understand the microbial life present in their fields and how it might affect crop health and yield.

What's unique about Trace Genomics is that whereas most soil tests look only at chemistry, the company also considers the biology of the soil. Farmers send soil samples to Trace Genomics and the company, then extracts DNA from these samples and sequences it to identify both beneficial and pathogenic microbes.

The massive amounts of sequencing data are then processed using machine learning algorithms. These algorithms are designed to identify patterns, correlations, and insights that would be hard for humans to discern from raw data. For instance, Trace Genomics can correlate the presence of certain microbes with crop yields and disease resistance.

Once the analysis is done, the company provides farmers with detailed reports that not only identify the microbes in their soil but also offer actionable insights. These insights could range from recommendations on crop rotation to the use of specific fertilizers to innovative ways to combat soilborne diseases.

As more and more soil samples are analyzed, the machine learning models become smarter. They can identify patterns and correlations with greater accuracy, making the recommendations more precise over time.

> Trace Genomics' AI-based insights range from recommendations on crop rotation to the use of specific fertilizers to innovative ways to combat soilborne diseases.

To provide a hypothetical example, let's say there is a farmer who has been facing declining yields in his cornfields over the past few years. Unsure of what's causing the decline, the farmer would send a soil sample to the company.

The analysis might reveal a high concentration of a certain pathogenic fungus that's known to attack corn roots. Additionally, the soil could have a lower-than-average concentration of beneficial microbes that help in nitrogen fixation. Based on these insights, Trace Genomics would recommend a specific biofertilizer that could introduce beneficial microbes to combat the pathogenic ones. The company might also suggest rotating corn with soybeans next year as soybeans can improve the soil's nitrogen content and suppress the harmful fungus.

AI-ENABLED LIVESTOCK MANAGEMENT AND ANIMAL HEALTH MONITORING

Farmers and agribusinesses are increasingly adopting AI-enabled livestock management and animal health monitoring to improve efficiency, reduce costs, and enhance the well-being of their livestock.

AI technologies are employed to analyze large volumes of data from various sources, including sensors, cameras, and wearable devices, in order to provide real-time insights into the health and behavior of the animals. Specific examples of how AI is being used in livestock are as follows:

1. Precision Feeding and Nutrition Management: AI algorithms can analyze data from various sensors and wearable devices attached to animals to track their feeding habits, weight, and growth rates. One common example is swine nutrition optimization. By combining this data with nutritional requirements and growth models, the AI system can optimize a pigs' diet in real time, ensuring it receives the right amount of feed with the appropriate nutrients. This precision feeding approach helps reduce feed wastage, enhances growth rates, and minimizes the environmental impact of animal agriculture.

2. Heat Stress Management: AI-based weather forecasting models can consider factors such as temperature, humidity,

and wind speed to forecast critical heat conditions. Armed with this information, farmers can take preventive measures like adjusting ventilation, providing cooling mechanisms, or modifying an animal's schedule to reduce the impact of heat stress on its health and productivity. One common example is using these AI models to predict upcoming heat stress events for poultry farms.

3. Reproduction and Breeding Optimization: AI can analyze data from wearable devices attached to dairy cows, such as activity monitors and health sensors, to predict the best time for artificial insemination based on estrus detection. By accurately identifying the optimal breeding window, farmers can enhance breeding success rates and improve the genetic traits of the herd.

4. Livestock Behavior Analysis: AI-enabled cameras can track the behavior of cattle, cows, or chickens and analyze patterns related to feeding, drinking, and movement. By monitoring behavioral changes, such as reduced feed consumption or abnormal movements, the system can alert the farmer to potential health issues, allowing for early intervention and reduced mortality rates.

5. Livestock Tracking and Identification: AI-powered tracking systems, such as radio frequency identification (RFID) tags and GPS collars, can monitor the location and movement of individual animals within a grazing area. This information helps farmers optimize pasture utilization, prevent overgrazing, and identify potential issues like straying or theft.

6. Cattle Health Monitoring: AI-powered systems equipped with computer vision and machine learning algorithms can analyze video feeds from cameras installed in cattle barns. These systems monitor the behavior and movements of individual

cows to detect any signs of distress, lameness, or abnormal behavior that could indicate health issues. For instance, if a cow is showing signs of limping or restlessness, the AI system can trigger an alert to the farmer, allowing them to promptly attend to the animal's health needs and potentially prevent the spread of disease.

Case Study: Stein Farms

Stein Farms is a 1000-herd operation located in Salem, Wisconsin, and in 2022, it deployed SmaXTec's TruRumi technology—a unique bolus technology that leverages data and artificial intelligence. The farm benefitted in a number of ways. First, after inserting the bolus in the cows, farmhands could identify mastitis in cows before they could see any of the symptoms. This early detection yielded significant savings by reducing the amount of antibiotics used.

The system also helped Stein Farms identify low-calcium cows as well as cows that were drinking copious amounts of water (a serious indicator that the animal might have ketosis). This allowed the farm to take appropriate action sooner than was previously possible. The same technology can prove helpful in isolating cows that might have infectious diseases, thus preventing potential outbreaks.

The technology also helped Stein Farms breed heifers more efficiently. For example, the technology is very good at deciding which heifers need to go to a veterinarian for a pregnancy check and which simply need to be re-inseminated. The technology can also accurately detect those animals in heat—often with no visible signs—and can provide "calving alerts" 12 hours before a cow actually delivers her calf.

Another company using AI to detect health problems in livestock is CattleEye. The company has developed a proprietary system that uses drones, cameras, and AI solutions to detect atypical cattle behavior as well as determine the impact of diet on livestock. By automating the process of collecting data on animals (which reduces the subjective nature of some of the data collected), the company helps farmers and ranchers maintain optimal body condition for cattle and cows by eliminating manual collection.

Using AI, CattleEye can recognize individual animals based on their natural coat patterns and facial features. This eliminates the need for invasive hardware such as ear tags or collars for identification. CattleEye's AI models can also detect subtle changes in an animal's gait, allowing farmers to spot potential lameness issues early on.

AI can provide calving alerts 12 hours before a cow actually delivers her calf.

Other AI companies, such as Cowlar and Connecterra, provide state-of-the-art wearable devices for cows. The devices include collars equipped with sensors, such as rumination sensors, accelerometers for activity monitoring, and temperature sensors for health assessment. The data collected from these devices are wirelessly transmitted to a central database where machine learning algorithms analyze the sensor data and identify patterns related to cow health, medical history, breeding records, and milk production data.

Because the AI model continuously monitors the cows' activity levels, rumination patterns, and body temperature to identify deviations from normal behavior, it can detect early signs of illness, reduced rumination, or increased restlessness.

HOW AI IS TRANSFORMING IRRIGATION AND WATER PRACTICES IN FARMING

In recent years, the agriculture industry has witnessed a revolution in irrigation practices, thanks to the introduction of artificial intelligence. AI-driven technologies are enabling farmers to optimize their irrigation systems and improve crop yields while at the same time reducing water usage and costs.

AI-driven technologies are helping farmers monitor and manage their irrigation systems more efficiently. By collecting data from sensors and other sources, AI-driven technologies can detect and analyze changes in soil moisture levels, weather conditions, and other factors that affect crop growth. This data can then be used to adjust irrigation schedules, which allows farmers to apply water more precisely and efficiently. Furthermore, AI-driven technologies can help farmers reduce labor costs by automating certain irrigation tasks.

AI can also help farmers optimize their crop selection and planting strategies by analyzing data from weather forecasts, soil moisture sensors, and satellite imagery. Farmers would then be able to identify the best crops for their land.

Two additional water-related areas where AI is proving helpful to farmers are improving the efficiency of water pumps and monitoring water quality.

AI-powered pumps can now be programmed to adjust water pressure and flow based on soil moisture and other conditions, and AI-powered sensors can also detect contaminants in water sources. The former helps farmers use the right amount of water for their crops, reducing water waste, while the latter can alert farmers to potential water contamination and help them protect their crops and the environment from water pollution.

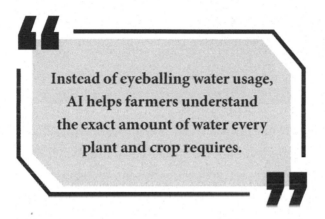

Instead of eyeballing water usage, AI helps farmers understand the exact amount of water every plant and crop requires.

Perhaps most importantly, AI is being used to optimize crop growth. AI-powered systems can analyze data from sensors and weather forecasts to determine the best times to irrigate crops. This helps farmers save water and increase crop yields.

Case Study:
Hortau

As the world's population continues to grow, so does the demand for food. This has put a strain on agricultural production, particularly in terms of water resources. In order to ensure crops are adequately supplied with water, farmers need to have access to reliable information about their water needs.

To address this challenge, a number of companies are leveraging artificial intelligence to monitor and predict water needs in agriculture. One such company is Hortau. Since 2003, its patented irrigation system has been helping farmers. At the time of this writing, Hortau was assisting over 1,300 farms and had over 300,000 acres under management, covering everything from almonds, cranberries, cherries, and grapes to alfalfa, onions, and wheat.

By combining real-time field data with weather predictions, Hortau's soil tension probes gauge the stress plants are under due to water availability. They can then report back to farmers' smartphones via a mobile web app. This helps farmers irrigate only when necessary, saving both money and water.

In one case study, a cherry farmer using Hortau's technology was able to discern that his west field was being exposed to greater sunlight than his east field. Instead of watering both fields 24 hours over a week's time, the farmer was able to increase his yield by applying 30 hours of water to the west field and only 18 hours to the east field.

In another case, a pistachio farmer came to understand that his land had radically different irrigation patterns and that applying water to those parcels that had gravel soil could be better served by a different irrigation pattern. As the farmer said, the biggest benefit was that they "were no longer eyeballing it." Instead, he was applying the right amount of water, at the right time, to the right crop.

> **According to the World Economic Forum, the integration of AI into agriculture could bring about a 60 percent decrease in pesticide usage and a 50 percent reduction in water usage.**

AI AND
REGENERATIVE
AGRICULTURE

As the world grapples with the challenges of climate change, food security, and environmental degradation, the need for sustainable farming practices is more crucial than ever. Regenerative agriculture, which focuses on improving soil health, biodiversity, ecosystem resilience, and sequestering carbon, offers a promising solution to these pressing issues.

AI-driven innovations are playing a crucial role in advancing regenerative agriculture, making it an attractive investment opportunity for those looking to support sustainable farming practices and contribute to a healthier planet.

One of the key principles of regenerative agriculture is the enhancement of soil health. Healthy soil is essential for growing nutritious crops, sequestering carbon, and maintaining the balance of ecosystems. AI technologies are being used to develop sophisticated soil monitoring systems that can provide farmers with real-time information about the health of their soil. These systems can also detect nutrient deficiencies, moisture levels, and other critical factors that affect soil health, enabling farmers to make informed decisions about the best practices for maintaining and improving their soil.

In addition to soil monitoring, AI is being used to optimize the use of resources in regenerative agriculture. For example, precision agriculture techniques are being combined with AI algorithms to create more efficient irrigation systems. These systems can analyze vast amounts of data to determine the optimal amount of water needed for each crop, reducing water waste and ensuring that plants receive the necessary nutrients for healthy growth. This not only conserves water resources but also helps to prevent soil degradation caused by overwatering.

Another area where AI is making a significant impact in regenerative agriculture is the development of alternate farming practices that promote biodiversity and ecosystem resilience. AI-driven technologies are being used to design and implement agroforestry systems, which involves the integration of trees and shrubs into agricultural landscapes. These systems can help increase biodiversity, improve soil health, and provide valuable ecosystem services, such as carbon sequestration and erosion control. Machine learning algorithms can also be used to analyze data on factors such as climate, soil type, and plant species to create customized agroforestry plans that maximize the benefits of these systems for individual farms.

AI is also being utilized to support the adoption of regenerative agriculture practices by providing farmers with access to information and resources. For instance, AI-powered chatbots and virtual assistants can be used to answer questions and provide guidance on topics such as soil health, crop rotation, and cover cropping. This can help bridge the knowledge gap that often exists between traditional farming practices and regenerative agriculture practices, making it easier for farmers to transition to more sustainable methods.

Case Study:
Indigo AG

Carbon by Indigo, a company specializing in agricultural carbon credits, supports farmers in their transition to more sustainable farming practices, including regenerative agriculture practices. Adding cover crops, reducing tillage, and rotating crops are just a few steps farmers can take to bring carbon back to their soil. Indigo currently has over 2000 farmers and 5.5 million acres enrolled.

Carbon by Indigo's system begins by helping farmers either add new practices or intensify existing practices. Next, soil samples and on-farm data are collected (and anonymized), with the results securely shared and verified by Indigo. The company then facilitates the payout process by connecting farmers with corporations who are looking to lower their carbon footprint and, thus, are willing to purchase carbon credits.

Under the existing system, farmers keep 75 percent of the carbon credits and Indigo retains the remaining 25 percent in return for scientifically measuring and verifying how much additional carbon each farm stores. The money a farmer receives will depend on a combination of the amount of additional carbon stored and the current market rate for carbon credits. It is not unreasonable to expect compensation in the range of $30 per acre.

AI AND INDOOR FARMING

Indoor farming, also known as controlled environment agriculture (CEA), refers to the process of producing food and medicinal plants within an indoor setting, whereby all aspects of the growing process are controlled. Indoor farming techniques include hydroponics, aeroponics, aquaponics, and vertical farming.

Indoor farming is expected to grow significantly in the years ahead—from an estimated $37 billion in sales in 2023 to between $90 and $100 billion by 2030—due to growing urbanization, changing consumer demands and tastes, and concerns over the sustainability of the existing global food supply chain.

The trend of indoor/urban farming is also expected to benefit from continued advances in LED lighting, hydroponics and aeroponics, robotics, and artificial intelligence. AI has the potential to significantly improve the efficiency, productivity, and sustainability of indoor farming operations. Below are ten specific ways AI is being utilized in indoor farming:

1. **Nutrient Management:** In hydroponic and aquaponic systems, AI can monitor and analyze the nutrient levels in water and adjust them as needed for each specific crop, optimizing inputs for both plant health and resource use. AeroFarms, an indoor farming company located in Newark, New Jersey, uses machine learning algorithms to analyze the collected data and adjust the

nutrient mix delivered to its crops via its aeroponic misting system. This allows the company to tailor the nutrient mix to the specific needs of each type of plant at different stages of its growth cycle.

2. **Climate Control Optimization:** AI algorithms can process data related to temperature, humidity, CO_2 levels from various sensors and adjust the indoor climate to optimize growing conditions for specific crops. Because AI systems continuously learn and improve, finding the perfect balance that encourages plant growth while minimizing water, nutrient, and energy consumption will soon be within the reach of indoor farmers.

> **Because AI systems continuously learn and improve, finding the perfect balance that encourages plant growth while minimizing water, nutrient, and energy consumption will soon be within the reach of indoor farmers.**

3. **Lighting Optimization:** Photosynthesis rates can vary based on the light spectrum. AI can modulate the light spectrum delivered by LEDs in real time based on the growth phase of the plant. This optimization can lead to faster growth cycles and energy savings.

4. **Consumer Feedback Loop:** By analyzing consumer feedback and purchase data, AI can provide insights into market demand, helping farmers adjust crop types and quantities.

5. **Integration with Smart Grids**: Large indoor farms can be integrated with smart grids, allowing them to adjust their energy consumption in response to wider grid demands, further optimizing energy costs.

6. **Space Utilization**: In vertical farms, space is a premium. AI can analyze growth patterns and adjust planting arrangements to maximize the use of available space. This helps ensure optimal light exposure and air circulation for every plant.

7. **Water Management**: One of the benefits of indoor farming is water conservation. AI can enhance this conservation by monitoring the moisture levels and ensuring plants receive water only when needed. AeroFarms, for example, uses a patented aeroponic system in which plant roots are misted with a nutrient-rich water system instead of being submerged in water or soil. This allows for a precise amount of water to be used on each plant and assists in determining the precise amount of water a plant needs at each stage of its growth. As an additional benefit, AeroFarms captures, cleans, and recycles all of the water it uses. This has resulted in growing crops with 95 percent less water than in traditional farming.

8. **Automated Harvesting:** AI-powered robots can be trained to recognize when a plant is ripe and ready for harvest, and they can then pick the produce without damaging the plant. This can save labor costs and increase efficiency.

9. **Crop Growth Prediction**: By analyzing data from previous growth cycles, AI can predict crop yields with increasing accuracy, allowing farmers to plan harvesting and sales more effectively.

10. **Seed to Sale Traceability**: AI can help in automating the tracking of produce from seeding, growth, harvest, to the point of sale. This can help in quality assurance as well

as understanding the overall efficiency of the farm. Such information will also prove helpful in the event of a recall by rapidly identifying the original source of the problem.

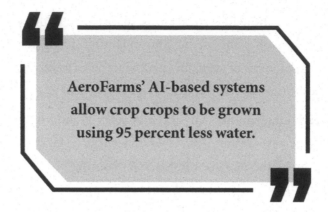

AeroFarms' AI-based systems allow crop crops to be grown using 95 percent less water.

Case Study:
Plenty

Plenty is a US-based agriculture technology company specializing in indoor vertical farming. It seeks to produce fresh, high-quality, and sustainably grown produce near urban centers using less land and water than traditional agriculture. The company starts by gathering vast amounts of data from sensors distributed throughout its farms. These sensors monitor numerous variables, including temperature, humidity, CO_2 levels, nutrient levels, and light intensity. The end result is a rich dataset that provides detailed insights into the growing conditions for each plant.

Plenty then uses machine learning algorithms to analyze the data it collects and gain insights into the optimal conditions for different crops at various stages of their growth cycles. By training these algorithms on historical data, Plenty's systems can predict the ideal conditions that will result in the best yield, flavor, texture, and nutritional content. By precisely controlling and optimizing the growing conditions, Plenty also aims to minimize its use of resources like water and nutrients.

> **Plenty's systems can predict the ideal conditions that will result in the best yield, flavor, texture, and nutritional content.**

Based on these insights, Plenty employs its automated systems to adjust the environment inside its farms in real time. These systems can alter the light spectrum, intensity, and duration to best match the needs of the crops at each stage of their growth. They can also adjust the nutrient mix, temperature, and humidity levels, thus tailoring the environment to the needs of the specific crops being grown.

Plenty also integrates advanced robotics into its operations. These automated systems, informed by AI analysis, handle tasks such as planting, harvesting, and packaging. This minimizes the potential for human error and helps reduce labor costs.

Plenty is also using AI in the digital tracking of seeds. From the moment a seed enters the facility, every aspect of the seed—variety, source, date of arrival, associated quality control data, etc.—is logged into its digital tracking system. This data creates the starting point of the traceability chain. Ultimately, consumers can scan a QR code on the packaging to see all the data associated with any batch of produce—from the seed it originated from, through its growth and harvest, to its journey to the store. This transparency can help to build trust with consumers. In the event of a recall or safety concern, Plenty's seed-to-sale traceability system allows the company to quickly and precisely identify which batches are affected. This ensures that only the necessary products are recalled, minimizing waste and economic loss while also protecting consumer safety.

FOOD QUALITY CONTROL AND AI

With the advent of AI, farmers, agribusinesses, and food companies are now able to use cameras, sensors, and sophisticated machine learning algorithms to improve the productivity and quality of agricultural products. Specifically, AI can help with food quality control by addressing issues such as bruising, rotting, early signs of spoilage and damage, or the detection of foreign objects.

One country for farmers to keep their eye on is Israel. It is currently the home of approximately 500 different AgTech companies. For example, Tevel is currently employing sensors, 3D cameras, drones, and AI on farms in the U.S, Israel, Chile, and Italy.

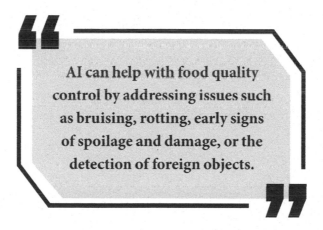

> AI can help with food quality control by addressing issues such as bruising, rotting, early signs of spoilage and damage, or the detection of foreign objects.

The company uses the cameras and AI to determine which fruit is ripe for picking, and then its drones pluck the fruit from the trees using soft robotic grippers that won't damage or bruise the fruit. In the process, the company's technology checks for diseases and can even measure sugar levels. Because Tevel's technology is completely autonomous, it is also helping address worker shortage issues.

BloomX, another Israeli company, uses drones and AI to help pollinate avocado trees. It does this by using electrostatic charges to mimic the movements of a bee. In one avocado farm, its technology increased yield by 40 percent.

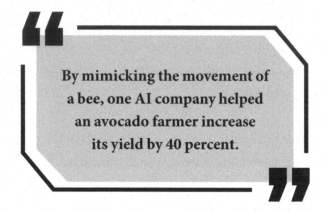

By mimicking the movement of a bee, one AI company helped an avocado farmer increase its yield by 40 percent.

Case Study:
Iron Ox

Iron Ox is a U.S.-based start-up seeking to address the challenges of modern agriculture, such as labor shortages, the need for sustainability, and the desire to produce food closer to the urban centers where it is consumed. (The last outcome, by allowing fresh produce to be grown as close to the consumer as possible, is expected to yield a significant reduction in the carbon footprint associated with transporting food.) To achieve these goals, the company is focused on developing autonomous robotic systems for indoor and urban farming.

Iron Ox's robotic systems handle the tasks of planting, caring for, and harvesting crops. The company employs advanced computer vision systems to monitor the health of the plants, and its systems can identify issues such as disease, pests, and nutritional deficiencies in real time. Based on this information, robots can be directed to take specific actions, such as adjusting environmental conditions or selectively treating individual plants. The company's goal is to produce high-quality food which is optimized for freshness and nutrition.

HOW CAN AI BE USED BY FARMERS TO IMPROVE FARM MANAGEMENT OPERATIONS?

In addition to the areas already mentioned in this book where AI can be beneficial to farmers, including precision agriculture, disease and pest control, weather prediction, livestock monitoring, soil health, and irrigation, AI can also help with market demand prediction, post-harvest sorting and packing, financial planning, and crop insurance.

Agribuddy, for example, is harnessing AI to provide farmers with market trends and pricing insight. Ripe.io is using blockchain technology and artificial intelligence to improve transparency in food quality, safety, and origin across the entire supply chain process, and Climate Corporation offers farms of all sizes a platform that uses AI to provide them with insights into their operations and offers insurance products based upon those insights.

Case Study: ClimateAI

As weather patterns become more volatile, the traditional methods farmers have used to make predictions and forecasts on what crops to plant are becoming less helpful. ClimateAI created its "Climate Lens" platform to combine AI, advanced machine learning, and data points from multiple sources in order to generate actionable insights designed to help farmers "future-proof" their operations.

In one case, ClimateAI used its AI-based predictive models to access high-quality satellite imagery as well as data about the ocean's temperature to improve its long-term forecasts. These insights were used by farmers, agribusinesses, and governments around the world to better assess when rice was at risk in a specific geographic location. It also used the data to determine an emerging opportunity for farmers. In this case, Climate AI's insights allowed rice farmers to make a successful transition to growing millet.

THE BIG BOYS ARE FOCUSING ON AI APPLICATIONS IN FARMING AND AGRICULTURE

There are hundreds of AI start-ups now competing in the agricultural arena. It is impractical to list every company, especially as some will be acquired, go out of business, or pivot to new areas in the months and years ahead. Moreover, new start-ups with innovative AI-based approaches, business models, and technology, will undoubtedly spring to life. One thing in this environment that is pretty clear is that the big companies will stay involved and continue to apply AI-based solutions to agriculture. In the past few years, a number of Fortune 100 companies have begun focusing on this emerging opportunity. Below is a short overview of some of the larger companies involved in AI.

Microsoft: FarmBeats is one of Microsoft's most notable contributions to agriculture. The project aims to enable data-driven farming by leveraging low-cost sensors, drones, and machine learning algorithms. By collecting and analyzing data from various sources, FarmBeats helps farmers make more informed decisions about their crops. For instance, by analyzing soil moisture data, a farmer can decide when and where to irrigate and, in the process, significantly reduce water wastage.

Microsoft's "AI for Earth" initiative provides grants to organizations and individuals working on environmental challenges across four domains: agriculture, biodiversity, climate change, and water. In the agriculture domain, AI for Earth supports projects that utilize AI to develop new farming practices and protect crops from pests. Recently, the company has been working with Bayer (and Climate Corporation) to link its data sources with Microsoft's Azure cloud platform.

Microsoft is also applying AI to its Project Premonition, an initiative designed to detect pathogens in the environment before they cause outbreaks harmful to people and animals.

Alphabet: Project Mineral is part of Google X—the moonshot factory of Alphabet (Google's parent company). Unveiled in 2020, Project Mineral focuses on sustainable food production and agriculture. Given the world's growing population and the changing climate, ensuring adequate food production is a pressing challenge, and Project Mineral is leveraging advanced technologies, including AI, to tackle this challenge.

One of the better-known tools from Project Mineral is the "plant buggy." These buggies roll through fields and, using a range of sensors, collect high-quality, real-time data about crops from the perspective of the plant (i.e., how plants grow, compete for resources, and respond to stress)

By using this data and applying its proprietary machine learning algorithms, Project Mineral seeks to identify patterns, make predictions, and provide actionable insights to farmers. Project Mineral also hopes to help plant breeders develop new varieties of crops. With the detailed data provided by the project, breeders can get insights into which traits are most beneficial under certain conditions and can breed crops that are more resilient, productive, and sustainable.

John Deere: One of the world's leading agricultural machinery companies, John Deere has been at the forefront of integrating AI and other advanced technologies into its products and solutions to drive the next revolution in agriculture.

In 2017, John Deere acquired Blue River Technology, a company specializing in creating smart machinery for agriculture. One of its more notable products is its "See & Spray" technology, which uses computer vision to detect and spray weeds in real time. This ensures herbicides are only applied where needed. In 2023, the technology became operational in its "ExactShot" system, which promises to reduce fertilizer use by as much as 60 percent.

> **In 2023, John Deere debuted its new planting technology "ExactShot" became available and it promised to reduce fertilizer use by as much as 60 percent by only using inputs when and where necessary.**

John Deere has also integrated AI into its machinery to enable features like autonomous driving. Its tractors and combines can automatically adjust their paths in the field to ensure maximum efficiency and minimal overlap. This reduces fuel consumption, decreases the wear and tear on machines, and ensures consistent operation.

Another area where John Deere is leveraging AI is predictive maintenance. Using AI and IoT sensors, the company's machinery can predict when parts are likely to fail or when maintenance is required. This predictive approach reduces downtime and helps in the seamless operation of farming activities.

IBM: The company's "Watson Decision Platform for Agriculture" merges AI, weather data, sensors, and blockchain to give farmers real-time actionable recommendations. Among other things, IBM uses machine learning to analyze weather data (from its subsidiary, The Weather Company) and identifies conditions that could lead to disease or pest infestation, allowing farmers to take preventive actions. IBM also helps farmers with yield forecasting by providing estimates based on current conditions, historical data, and AI analysis. It's AgroPad, for example, allows farmers to analyze soil or water samples using a paper testing strip. When this strip is photographed with a smartphone, the companion app, powered by machine learning, provides a chemical profile within seconds.

Cargill: As America's largest privately held global food corporation, Cargill is investing aggressively in artificial intelligence. Its Galleon Microbiome Analysis is a comprehensive microbiome health assessment tool that allows broiler producers to understand how changes in raw materials, diet, additives, vaccine programs, and farm management practices can influence the microbiome of their flock. Using a simple swab from a live bird, Galleon compares the bird's microbiome against a large and growing database of poultry microbiome and can help tell producers if they need to switch feed or additives.

Cargill has another AI tool, Birdoo, which leverages proprietary computer visioning technology and AI to provide hands-free, real-time insights on the health of flocks. In addition to saving on labor costs, the technology replaces manual weighing and allows for better harvest planning by reducing variability and cutting down on the number of downgrades that can occur at the process packing plant.

CHALLENGES AND BARRIERS REGARDING AI AND FARMING

Knowing when to invest in any technology is a complicated and imperfect process. This is especially true in farming, where the vagaries of weather, changing market conditions, ever-shifting consumer preferences, and political and regulatory issues are ever present and create a high level of uncertainty. Investing in artificial intelligence is no different from any other decision regarding technology except for the fact that it might be even less straightforward than an investment in a tractor or other piece of equipment.

It is also entirely possible that now is not the time to "pull the trigger" on artificial intelligence. As was mentioned earlier, the society is still in the earliest stages of this revolution, and advances in AI are likely to be usurped by next-generation advances that may arrive next year or even next month.

One analogy to keep in mind is that of the telephone. Most farms had a landline by the 1940s and this technology stayed the same until about 2000. Then the first mobile phones were developed but they were bulky in size, expensive, and prone to not working in all conditions. Next came flip phones and Blackberries which were slightly more affordable and had some new features because they could also access the Internet. Eventually, the smartphone, as we currently know it, arrived.

In this same way, the early advances in AI will be good and maybe even helpful but understand they will get better quickly. The best way to consider making an investment is to consider return on investment. This will differ for every farmer and every business but, as a general rule, if an AI solution can pay for itself over a two-to-three-year period, it is worth considering an investment. (At a minimum, investing $20 a month to gain access to ChatGPT4—and, soon, ChatGPT5—is a low-risk, prudent decision.)

However, before pulling the trigger, here are a few things to consider:

1. **Large upfront costs**: While AI solutions can be cost-effective in the long run, the initial investment can be expensive. Before entering into any partnership, beware of the upfront cost. Also seek government grants and look for partners who are willing to entertain "win-win" scenarios such that they profit only when you profit.

2. **Data**: A second challenge involves the collection and processing of data. For AI algorithms to operate effectively and for machine learning to realize its potential, the systems require large volumes of high-quality data that can be used to train the models and improve their accuracy over time. Since this requires a sizable investment in sensors, satellite data, equipment, and labor, it may exceed the financial means of many farmers.

3. **Connectivity**. Internet access is another key challenge. While fiber optics and satellite Internet platforms continue to get better and drop in price, many farmers still have limited or unreliable Internet connectivity.

4. **Privacy and Security**. While not unique to AI, cyberattacks, data leaks, and questions over who owns and controls the data and the insights need to be considered before making any decision regarding AI and which vendors to use.

It is only natural that people are hesitant to adopt new technologies. This is especially true in agriculture where the average age of a farmer in the United States is 58. One way to think of AI is to understand that at its most basic level, AI is only a more advanced version of field data processing. Of course it is impossible to know beforehand if the insights uncovered will cover the costs of investing in new technologies or systems. Farmers can minimize the downsides of AI by speaking with other farmers currently employing the technology and by doing their due diligence.

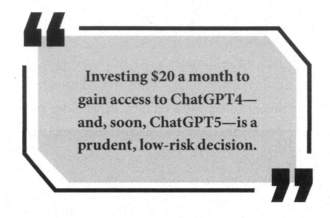

Investing $20 a month to gain access to ChatGPT4— and, soon, ChatGPT5—is a prudent, low-risk decision.

LEGAL IMPLICATIONS OF AI AND FARMING

The integration of AI into farming and agriculture brings forth several legal implications. As with any emerging technology, regulation is required to strike a balance between innovation and potential risks. Some of the legal considerations concerning AI and farming are as follows:

1. **Data Privacy and Ownership**: One of the biggest challenges is determining who owns the data collected from the farm, especially when AI systems are offered by third-party vendors. There are concerns about how this data might be used, particularly if it's shared with other entities or used to influence market prices. Regulations like the General Data Protection Regulation (GDPR) in Europe offer some guidelines, but the specific context of farming might require more tailored rules. Before engaging with any third party vendor, farmers should understand who owns the data and how and with whom it might be shared.

2. **Liability**: If an AI-driven system makes a decision that results in crop failure or another significant setback, it raises the question of who is liable—the farmer, the AI developer, or the technology provider. This becomes even more complicated with autonomous machinery.

3. **Intellectual Property**: There may be issues around who owns the algorithms, models, and insights generated by AI systems in farming. As companies develop proprietary AI models tailored for specific agricultural tasks, IP rights will become crucial.

4. **Safety and Standards**: With the potential use of autonomous farming equipment like tractors or drones, there need to be safety standards in place. Who is responsible if a machine causes damage or injury? And what safety benchmarks do these machines need to meet?

5. **Transparency and Trust:** When it comes to food, people want transparency in how it's produced. If AI is making decisions, farmers and consumers alike must understand how these decisions are being made and, if necessary, challenge the "black box" nature of some AI algorithms. (In simpler terms, AI must be able to explain to its users how it arrives at its answers.)

6. **Environmental Concerns**: Because AI systems rely on an enormous amount of computing power, the electricity and energy costs may pose serious concerns. AI, in its current configuration, has a relatively high carbon footprint.

7. **Labor and Employment**: As AI systems replace tasks traditionally done by farmworkers, there can be implications for farm employment. Regulations may be needed to ensure a fair transition for displaced workers.

8. **Ethical Treatment of Livestock**: If AI systems are used in livestock management, there may be concerns and regulations around the ethical treatment of animals.

9. **Trade and Market Implications**: AI can provide farmers or agribusinesses a significant advantage in terms of yield prediction, market analysis, etc. This might have implications for trade agreements and/or market competition regulations.

10. **Access and Equity**: There's a risk that the benefits of AI in farming could be limited to those who can afford the technology, thereby increasing the divide between large, commercial farms and smaller farms. Regulations may be required to ensure equitable access to AI benefits for all farmers.

As AI continues to be integrated into farming, ongoing dialogue between stakeholders, including farmers, technologists, legal experts, and policymakers will be crucial in navigating these and other legal implications.

RESOURCES

In May of 2023, the National Science Foundation announced the creation of seven new artificial intelligence research institutes for the purpose of harnessing the opportunities surrounding AI and addressing the risk associated with the technology.

The initiatives, which are funded with $140 million, cover a wide spectrum of fields and activities. The two involving agriculture are Washington State University's AgAid initiative and Iowa State University's AI Institute for Resilient Agriculture (AIIRA). The former seeks to employ AI methods to help farmers address challenges related to labor, water, weather, and climate change. The latter will focus on transforming agriculture through innovative AI-driven digital "twins" of individual plants and entire farm fields.

> **AIIRA seeks to transform agriculture through innovative AI-driven digital "twins" of individual plants and entire farm fields ... meaning farmers will be able to run unlimited 'what-if' scenarios in order to make better decisions in the real world.**

AIIRA researchers will continuously feed real-time data from sensors in the biological plants and fields—including weather data, soil measurements of water and nitrogen, soil and topography maps, ground and drone-based imaging, and satellite information—into predictive digital twins in order to create a functional, virtual representation of either a real plant or a real farm.

With this information, AIIRA researchers and farmers will be able to run unlimited "what-if" scenarios in order to make better decisions in the real world. The program aims to give scientists better insights into plant genetics and physiology for the purpose of breeding plants that can better adapt to changing environments and emerging diseases and pests. It also aims to help farmers optimize planting decisions and to make the most efficient decision regarding fertilizers, irrigation, and pesticides.

The two institutes hope to research the potential of AI to deliver broad societal and economic benefits as well as push the boundaries of what AI can achieve by training students, educators, and the workforce in AI technologies and their applications.

HOW CAN A FARMER BEST STAY ABREAST OF ADVANCES IN AI?

A s was stated at the beginning, due to the accelerating pace of advances in artificial intelligence, this book is already outdated. It is important, therefore, that farmers continue to stay abreast of advances in artificial intelligence. The following is a list of possible strategies:

1. **Subscribe to my weekly newsletter,** *The Friday Future Five.* The newsletter is free and readers can subscribe at <u>www.jackuldrich.</u> com. Every week, I provide five articles which I encourage subscribers to read and think about. While it is not guaranteed that every week will contain an article specifically related to agriculture and AI, I do my best to follow advances in the field and share the highlights on a consistent basis. Matt Wolfe's "Future Tools Weekly" is also an excellent free resource on AI.

2. **Attend online courses and workshops**. There are many online platforms including Coursera, edX, Udemy, MIT, and more that offer courses in AI. Some are general introductions to the technology, while others focus on specific applications in agriculture.

3. **Read Industry publications**: Journals, magazines, and online articles that are focused on agriculture and technology can also provide updates on the latest trends, breakthroughs, and

case studies. While researching for this book, I found MIT's Technology Review, ScienceDirect, Fast Company, Precision Farm Dealer, and Farm Progress to be helpful in monitoring advances in farming and artificial intelligence.

4. **Attend conferences and seminars**: Attending or following conferences focused on agriculture technology can provide insights into the latest research and development in AI for farming. Events like the Precision Agriculture Conference, InfoAg, AI Institute for Next Generation Food Systems, Food and Agriculture Organization's "AI for Good Global Summit," IEEE's Symposium on Computational Intelligence for Agriculture, and World Agri-Tech Innovation Summit are good places to start.

5. **Engage with local universities, agricultural extension offices, and cooperative**s: Many universities are at the forefront of AI research and can provide practical insights. Stanford, University of California at Davis, Cornell, Purdue, Iowa State University, and the University of Illinois are among the leading American institutions focusing on artificial intelligence and farming. Consider collaborating with or attending seminars at those universities near you. A number of leading cooperatives, including Land O'Lakes, CHS and Landus are also doing an outstanding job of helping their members understand how to best leverage AI.

6. **Network with like-minded organizations**: Join AI and AgTech groups or online forums. Sites like LinkedIn have various professional groups dedicated to the intersection of AI and agriculture. These not only provide information but also facilitate connections with experts in the field.

7. **Seek out AI solution providers**: If you're already working with a company that provides AI solutions or tools for farming—be it Farmers Business Network, John Deere, Cargill, Bayer, or any

of the hundreds of other agricultural AI start-ups—regularly communicate with them about new features and advancements. These companies often have a vested interest in keeping their customers informed. Another idea is to consider becoming a test farm for tech start-ups in the AgTech/AI space. This can provide early access to the latest tools and innovations.

8. **Listen to Podcasts and view webinars**: There are several technology and agriculture-focused podcasts and webinars covering recent advancements in AI, including "Last Week in AI," the "TWIML AI Podcast," and "Your Breakdown of AI News for the Past Week." These podcasts are a convenient way to get updates while on the go.

9. **Seek out Government resources**: Governments around the world are increasingly focusing on digital agriculture. They may provide resources, training sessions, and updates on AI applications in farming.

10. **Visit demonstration farms**. Some regions have demonstration farms where new technologies, including AI, are being tested and showcased. Visiting these farms or collaborating with them can offer hands-on experience. For farmers in the Midwest, one such location is North Dakota's Emerging Prairie, which is working with both large and small AI providers.

11. **Participate in field trials**: If feasible, run small-scale trials on your farm. This will allow you to test the effectiveness and utility of AI tools specific to your conditions.

12. **Stay curious**: As with any rapidly evolving field, having a sense of curiosity, open-mindedness, and a willingness to adapt is crucial. Regularly seek out information and be open to testing and integrating new methodologies into your farming practices.

CONCLUSION

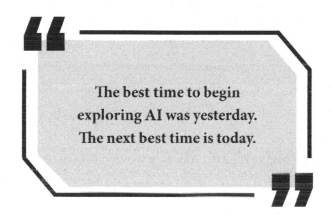

The best time to begin exploring AI was yesterday. The next best time is today.

Technology has always been at the forefront of agriculture and, for the foreseeable future, AI will play a pivotal role in agriculture and food sustainability. Whether it is monitoring crops and improving our understanding of how plants grow and respond to changing environments, minimizing the use of water and fertilizers, or enhancing food production at scale without negatively impacting natural resources, AI is poised to revolutionize modern agriculture by improving the efficiency, sustainability, and production of healthier and higher-quality crops and produce.

Smart farming tools, intelligent automation, and AI-powered products will perform repetitive, time-consuming tasks so workers can use their time for more strategic operations which require human judgment. Furthermore, increasingly affordable computer vision tools working alongside agricultural robotics and drones will continue to accelerate AI advancement in farming.

What this means is that AI is likely to change the role of farmers from being manual workers to becoming the planners and overseers of smart agricultural systems. It is important, therefore, to stay abreast of how artificial intelligence will continue to transform farming. It is not too soon to begin reaching out to your local universities, cooperatives, and leading agribusinesses to understand how they are thinking about and utilizing AI today.

The success of human civilization is essentially dependent on the optimization of its agricultural systems. Traditional farming methods are becoming outdated and there is a growing need for advanced technological solutions. Worldwide, the impact of automation on industries has always been considerable. Digital technology is now playing a huge role in transforming agriculture, and the impact of artificial intelligence in agriculture is set to be vast.

The best time to begin exploring AI was yesterday. The next best time is today.